ULTIMATE GUIDE TO BED BUGS-FREE HOME

Comprehensive Strategies For Prevention, Treatment, And Control Of Bed Bug Infestations: Detection, Identification, Removal, Safety, And Effective Solutions

Ethan Harry

Table of Contents

CHAPTER ONE 4

UNDERSTANDING BED BUGS 4

Bed Bug Biology And Behavior..7

Common Myths About Bed Bugs 10

Identifying Bed Bug Infestations..12

CHAPTER TWO 16

PREVENTION STRATEGIES 16

Keeping Your Home Clean And Clutter-Free.........................16

Inspecting Second-Hand Furniture And Clothing.................18

Using Protective Covers For Mattresses And Box Springs 21

CHAPTER THREE..25

HOME MAINTENANCE AND REPAIRS 25

Sealing Cracks And Crevices..25

Repairing Wall Damage And Loose Wallpaper 27

Proper Ventilation And Moisture Control..............................30

CHAPTER FOUR 33

TRAVEL PRECAUTIONS 33

Inspecting Hotel Rooms 33

Keeping Luggage Bed Bug-Free...35

Post-Travel Inspection And Cleaning 38

CHAPTER FIVE 41

EARLY DETECTION TECHNIQUES 41

Regular Home Inspections...41

Using Bed Bug Interceptors And Monitors 44

Recognizing Early Signs Of Infestation 46

CHAPTER SIX 50

SAFE AND EFFECTIVE TREATMENTS 50

Non-Chemical Treatment Options 50

Chemical Treatments And Safety 53

Professional Pest Control Services 56

CHAPTER SEVEN 60

LONG-TERM BED BUG MANAGEMENT 60

Developing A Bed Bug Prevention Plan 60

Educating Family And Household Members 62

Regular Follow-Up And Monitoring 65

THE END 68

CHAPTER ONE

UNDERSTANDING BED BUGS

Bed bugs are small, flat, oval-shaped insects that feed on the blood of humans and animals. They are reddish-brown in color and can be as small as a poppy seed when they are young, growing to about the size of an apple seed as adults. Bed bugs are wingless and have six legs. Despite their tiny size, they are visible to the naked eye.

These pests are notorious for infesting places where people sleep, such as homes, hotels, and shelters. Bed bugs typically hide in cracks and crevices during the day and come out at night to feed. They are often found in the seams of mattresses, box springs, bed frames, and headboards. They can also hide in furniture, behind wallpaper, and under carpets.

Bed bugs are not known to spread diseases, but their bites can cause discomfort and other health issues. The bites often result in red, itchy welts on the skin. Some people may develop allergic reactions to bed bug bites, leading to more severe symptoms like swelling and blisters. The bites are usually found in clusters or straight lines on the skin, often on exposed areas such as the face, neck, arms, and hands.

One of the most challenging aspects of dealing with bed bugs is their ability to spread quickly and their resilience. They can travel from one place to another by hitching a ride on luggage, clothing, and used furniture. Bed bugs are also known for being tough to get rid of. They can survive for several months without feeding, making them hard to eliminate through starvation.

To detect bed bugs, look for signs such as tiny blood stains on sheets, dark spots of bed bug excrement, shed skins, and a musty odor. If you suspect a bed bug infestation, it is crucial to act quickly. Cleaning and vacuuming infested areas can help, but professional pest control services are often necessary for complete eradication.

Preventing bed bug infestations involves regular inspections of sleeping areas, keeping luggage off the floor in hotels, and being cautious with second-hand furniture. By being vigilant and taking immediate action, you can reduce the risk of bed bug infestations and protect your home and health from these persistent pests.

Bed Bug Biology And Behavior

Bed bugs are small, parasitic insects that feed on the blood of humans and animals, belonging to the family Cimicidae. The most common species is Cimex lectularius. Adult bed bugs are oval-shaped, flat, and reddish-brown, measuring about 4 to 5 millimeters in length. Their flattened bodies enable them to hide in tiny crevices and cracks, making detection and elimination challenging.

Bed bugs are nocturnal creatures, most active at night. They are attracted to warmth and the carbon dioxide exhaled by humans and animals, which leads them to their hosts while they sleep. Bed bugs pierce the skin with two hollow tubes: one tube injects saliva containing an anticoagulant to keep the blood flowing, while the other tube withdraws blood. This feeding process typically takes several minutes.

A female bed bug can lay hundreds of eggs in her lifetime, often in hidden locations like mattress seams, furniture crevices, and wall cracks. These eggs are tiny, white, and sticky, making them hard to see and remove. Bed bugs undergo incomplete metamorphosis, with three stages: egg, nymph, and adult. Nymphs resemble smaller versions of adults and require a blood meal to molt and grow through five instar stages before reaching maturity.

Bed bugs are resilient and can survive for months without feeding, adding to the difficulty of eradicating them. They can endure a range of temperatures and environmental conditions. Their ability to hide and survive without food for extended periods makes it easy for them to spread and re-establish infestations after control measures are implemented.

Effective bed bug control requires a thorough understanding of their biology and behavior. Detection often involves inspecting common hiding places, such as mattress seams, furniture joints, and baseboards. Treatments typically include a combination of chemical and non-chemical methods, such as vacuuming, steam cleaning, and applying insecticides. Preventive measures, like encasing mattresses and reducing clutter, can also help manage infestations. Due to their resilience and ability to hide, bed bugs remain a challenging pest to control, necessitating persistent and integrated pest management approaches.

Common Myths About Bed Bugs

There are several misconceptions about bed bugs that contribute to the difficulty in dealing with them effectively. Here are some common myths and the truths behind them:

Myth 1: Bed bugs are too small to be seen with the naked eye. Truth: While bed bugs are small, they are visible to the naked eye. Adult bed bugs are about the size of an apple seed and can be seen without magnification. Nymphs and eggs, however, are much smaller and may be more challenging to spot.

Myth 2: Bed bugs only infest dirty or unkempt homes. Truth: Bed bugs are not attracted to dirt or filth; they are attracted to warmth and blood. They can infest any environment, regardless of cleanliness. Hotels, hospitals, public transportation, and even well-maintained homes can harbor bed bugs.

Myth 3: Bed bugs transmit diseases. Truth: While bed bugs can cause discomfort and secondary infections due to scratching, there is no scientific evidence that they transmit diseases to humans. Their bites can lead to allergic reactions in some individuals, but they are not known to spread pathogens.

Myth 4: Bed bugs are only found in beds. Truth: Although bed bugs are commonly found in beds and bedding, they can also infest other areas where people spend extended periods. These areas include sofas, chairs, carpets, baseboards, and even electrical outlets. Bed bugs will hide in any crack or crevice close to their food source.

Myth 5: You can get rid of bed bugs by discarding infested furniture. Truth: Simply discarding infested furniture is not a guaranteed solution, as bed bugs can spread to other areas of the home. Comprehensive treatment, including professional pest control services, is usually necessary to eradicate an infestation.

Identifying Bed Bug Infestations

Detecting a bed bug infestation early is crucial for effective control and minimizing the spread. Here are some key indicators to look for:

Bites: One of the most common signs of bed bugs is the presence of bites on the skin. These bites often appear as small, red, itchy welts, typically arranged in a line or cluster. However, it's important to note that not everyone reacts to bed bug bites, so the absence of bites does not necessarily mean there is no infestation. The reactions to bites can vary widely among individuals, ranging from no reaction at all to severe allergic responses.

Bloodstains: After feeding, bed bugs may leave small bloodstains on bed sheets, pillowcases, or mattresses. These stains are usually dark red or rusty in color and can occur when a bed bug is accidentally crushed after feeding. Finding unexplained bloodstains on your bedding is a strong indication that bed bugs may be present.

Fecal Spots: Bed bug excrement appears as small, dark spots or smears on bedding, walls, or furniture. These spots are digested blood and can look similar to mold or mildew. You may find these spots in areas where bed bugs hide, such as mattress seams, furniture joints, and baseboards. The fecal spots can also emit a distinctive, musty odor.

Shed Skins: As bed bugs grow, they molt and leave behind their exoskeletons. These shed skins can be found near their hiding places, such as mattress seams, furniture joints, or baseboards. The presence of shed skins is a strong indicator of an established infestation, as it shows that bed bugs are going through their life cycle in your home.

Live Bugs: Seeing live bed bugs is a clear sign of an infestation. Adult bed bugs are small, about the size of an apple seed, and are reddish-brown in color. Nymphs are smaller and lighter in color. Check common hiding places like mattress seams, box springs, headboards, and cracks in furniture. Bed bugs are nocturnal and tend to come out to feed at night, so inspecting these areas in the evening with a flashlight can be more effective.

Musty Odor: In severe infestations, bed bugs can produce a sweet, musty odor. This smell is often compared to that of coriander or moldy clothes. The odor comes from the bed bugs' scent glands and can be quite noticeable in heavily infested areas.

CHAPTER TWO

PREVENTION STRATEGIES

Keeping Your Home Clean And Clutter-Free

Keeping your home clean and clutter-free is one of the most effective ways to prevent bed bugs. These pests thrive in small crevices and cluttered environments, so minimizing their hiding spots is crucial. Here's how to make your home less appealing to bed bugs:

Regular Cleaning: Vacuum your home frequently, focusing on areas around the bed, furniture, and baseboards. Use a vacuum cleaner with a HEPA filter to ensure that any bed bugs or eggs are effectively captured. After vacuuming, immediately empty the vacuum cleaner outside your home to prevent any captured bed bugs from escaping back into your living space.

Reducing Clutter: Clutter provides bed bugs with more places to hide, making detection and elimination more challenging. Keep your home organized and minimize clutter as much as possible. Store items in sealed plastic bins rather than cardboard boxes, which bed bugs can easily infiltrate. This not only reduces hiding spots but also makes it easier to spot signs of bed bugs.

Frequent Laundry: Regularly washing and drying your bed linens, blankets, and clothes is essential. Use the hottest temperature settings that the fabric can withstand, as high heat can kill bed bugs and their eggs. Don't forget to clean items like pet bedding and stuffed animals, which can also harbor bed bugs. Frequent laundering of these items reduces the risk of bed bugs establishing themselves in your home.

Regular Inspections: Conduct routine inspections of your living space, especially in bedrooms and living rooms. Look for signs of bed bugs, such as small, rust-colored stains on bedding or tiny, dark spots (bed bug excrement) on mattresses and furniture. Early detection is key to preventing a full-blown infestation. Regularly inspect areas where bed bugs are likely to hide, including mattress seams, box springs, and behind headboards.

Inspecting Second-Hand Furniture And Clothing

Second-hand furniture and clothing can be significant sources of bed bug infestations.

While the allure of a great bargain is tempting, it's essential to conduct a thorough inspection to prevent these pests from entering your home.

Furniture Inspection: Before introducing any second-hand furniture into your home, a meticulous inspection is crucial. Begin by using a flashlight to examine the furniture's crevices, seams, and cracks where bed bugs are likely to hide. Upholstered furniture requires extra scrutiny as bed bugs can easily conceal themselves in fabric folds, seams, and cushions. Check all potential hiding spots, including underneath the furniture and inside any compartments. Look for signs such as small reddish-brown stains, dark spots of bed bug excrement, or live bugs. Ensuring the furniture is clean before bringing it inside can prevent a potential infestation.

Clothing Inspection: When buying second-hand clothing, it's important to inspect each item thoroughly. Examine seams, pockets, and folds in the fabric for any indications of bed bugs. Small eggs, nymphs, or even adult bed bugs can be hidden in these areas. To further ensure safety, wash and dry all second-hand clothing on the highest heat setting possible before wearing or storing them. The heat from washing and drying can kill bed bugs and their eggs, reducing the risk of an infestation.

Precautionary Treatment: To add an extra layer of protection, consider treating second-hand furniture and clothing with heat or a bed bug spray before bringing them into your home. Heat treatments, such as using a portable bed bug heater, are effective in killing both bed bugs and their eggs. If opting for a bed bug spray, select one that is specifically labeled for bed bugs and follow the manufacturer's instructions meticulously. Proper treatment can help ensure that any potential pests are eradicated before they have a chance to spread.

Using Protective Covers For Mattresses And Box Springs

Protective covers for mattresses and box springs offer a straightforward and highly effective method to guard against bed bug infestations. These covers act as a barrier, preventing bed bugs from infiltrating your mattress or box spring and establishing a home in these areas.

Choosing the Right Cover: When selecting a protective cover, prioritize those specifically designed to combat bed bugs. Look for products labeled as "encasements," which fully envelop the mattress and box spring and feature a secure zipper closure. These encasements should be constructed from durable materials capable of withstanding regular use and potential wear and tear. A high-quality cover will ensure that bed bugs cannot penetrate through the fabric, offering a solid line of defense against these pests.

Proper Installation: For the protective cover to be effective, proper installation is crucial. Begin by placing the cover over the mattress and box spring, ensuring it encases them completely without leaving any gaps or openings. Carefully zip the cover shut, paying close attention to the zipper to confirm it is securely fastened. Gaps or improperly sealed zippers can undermine the cover's effectiveness, allowing bed bugs to slip through and potentially infest your bedding.

Regular Inspection: Despite the protection offered by encasements, regular inspections are necessary to maintain their effectiveness. Periodically check your mattress and box spring for any signs of damage or wear, such as tears or holes in the cover. These imperfections could compromise the cover's ability to block bed bugs. Additionally, keep an eye out for any signs of bed bug activity, such as small dark spots or shed exoskeletons. If you discover any evidence of bed bugs, prompt action is essential. Treat the affected area and take steps to address the infestation immediately.

CHAPTER THREE
HOME MAINTENANCE AND REPAIRS

Sealing Cracks And Crevices

Preventing a bed bug infestation begins with addressing potential hiding places. Bed bugs are incredibly small and can squeeze into the tiniest gaps. To minimize the risk of an infestation, it's essential to thoroughly inspect your home for cracks and crevices that might serve as hiding spots for these pests.

Baseboards and Trim: Start by examining the areas where baseboards meet the wall. Gaps in these areas are common entry points for bed bugs. To address this, apply a bead of caulk along the seam between the baseboard and the wall. Smooth the caulk with a putty knife or your finger to ensure a tight seal, and check for any remaining holes or spaces that need to be filled.

Electrical Outlets and Switch Plates: Bed bugs can also hide behind electrical outlets and switch plates. Carefully remove these plates and inspect the space behind them. Any gaps around the outlet should be sealed with caulk or another suitable filler to prevent bed bugs from accessing these areas.

Window and Door Frames: Bed bugs may also find their way through gaps in window and door frames. Examine these frames closely for any cracks or gaps and seal them with weatherstripping or caulk. Ensuring that doors and windows close tightly is crucial to prevent bed bugs from entering or escaping. This can also help with general energy efficiency in your home.

Flooring and Carpets: Check the area where flooring meets the walls, as gaps in this area can be a potential hiding place. Seal these gaps with an appropriate sealant or filler. Ensure that carpets are securely tacked down and that there are no loose edges where bed bugs could potentially hide. Regularly inspect these areas to ensure that no new gaps or issues have developed.

Repairing Wall Damage And Loose Wallpaper

Maintaining your walls in good condition is crucial for preventing bed bugs from finding places to hide. Addressing wall damage and loose wallpaper can significantly reduce the risk of these pests settling into your home.

Wall Damage: Start by thoroughly inspecting your walls for any signs of damage, such as holes, cracks, or peeling paint. Small holes and cracks are common and can often be repaired easily. For minor holes, use spackling paste or joint compound. Apply the paste with a putty knife, smoothing it over the damaged area. Once it dries, sand it down until it's flush with the surrounding wall. For larger holes, you may need to patch the area with a piece of drywall, secure it with drywall tape, and then apply joint compound over the seams. Sand the area smooth once the compound is dry, and repaint as needed. This not only restores the appearance of your walls but also eliminates potential hiding spots for bed bugs.

Loose Wallpaper: Bed bugs are known to seek out hiding places, and loose or peeling wallpaper can provide an ideal refuge. Begin by carefully removing any loose or damaged wallpaper. Take care not to damage the underlying wall surface in the process. Once the wallpaper is removed, assess the wall for any additional repairs that may be needed, such as filling in holes or smoothing out uneven patches. After making the necessary repairs, you have two options: apply new wallpaper or repaint the wall. If you choose to reapply wallpaper, ensure that the edges are firmly adhered to the wall to avoid creating gaps where bed bugs could potentially hide. If you opt for repainting, select a high-quality paint and apply it evenly to provide a smooth, seamless finish.

Proper Ventilation And Moisture Control

Proper ventilation and moisture control are essential measures in preventing bed bug infestations. Bed bugs are notorious for thriving in warm, humid environments, so managing these conditions is crucial to keeping them at bay.

Ventilation is a key factor in maintaining a bed bug-free home. Effective ventilation helps keep indoor air fresh and reduces humidity, making it less inviting for bed bugs. To achieve proper ventilation, use exhaust fans in high-moisture areas such as bathrooms and kitchens. These fans help expel humid air and reduce overall moisture levels. Additionally, whenever possible, open windows to allow fresh air to circulate throughout your home. This not only improves air quality but also aids in controlling humidity levels. A well-ventilated home can significantly decrease the likelihood of bed bugs establishing a foothold.

Moisture control is equally important in preventing bed bug problems. Bed bugs are drawn to moist environments, so addressing any sources of excess moisture in your home is crucial. Start by identifying and fixing any leaks or damp spots. Plumbing leaks should be repaired immediately to prevent water from accumulating and creating a favorable environment for bed bugs. Ensure that your home's gutters and downspouts are functioning correctly to avoid water damage and moisture buildup. In areas prone to moisture, such as basements and crawl spaces, consider using a dehumidifier. Dehumidifiers help to reduce excess moisture in the air, keeping humidity levels low and making your home less attractive to bed bugs.

CHAPTER FOUR
TRAVEL PRECAUTIONS
Inspecting Hotel Rooms

Before settling into your hotel room, it's essential to conduct a thorough inspection to ensure it is free from bed bugs. Start by examining the bed area, including the mattress, box spring, and headboard, as bed bugs tend to stay close to their hosts. Look for tiny reddish-brown spots or small, dark stains that could indicate bed bug droppings.

Begin by checking the seams and edges of the mattress. Use a flashlight if necessary since bed bugs are nocturnal and may be challenging to spot in dim lighting. Carefully inspect the crevices of the box spring, another common hiding spot for these pests. Be meticulous in your search, as bed bugs are adept at hiding in small, dark places.

Next, examine the furniture around the bed. Nightstands and lamps can also harbor bed bugs. Pay special attention to any cracks or crevices where they might be hiding. Pull back the bedding and check the corners of the bed frame and headboard. Bed bugs can hide in these areas and come out at night to feed.

Don't forget to inspect other furniture in the room, such as chairs, couches, and dressers. Look behind picture frames and under any objects resting on flat surfaces. Bed bugs can hide in the smallest of spaces, so be thorough in your search.

If you find any signs of bed bugs during your inspection, immediately request a different room. Choose a room that is not adjacent to or directly above or below the infested room, as bed bugs can easily spread through walls and electrical outlets. If you find evidence of bed bugs in the new room, consider finding a different hotel altogether.

Avoid placing your luggage on the bed or floor before you complete your inspection. Instead, use a luggage rack away from walls and upholstered furniture. Once you are confident the room is free of bed bugs, you can unpack your belongings. By taking these precautionary steps, you can significantly reduce the risk of bringing bed bugs home with you from your travels.

Keeping Luggage Bed Bug-Free

Traveling can be an enjoyable experience, but the risk of bringing home unwelcome guests, such as bed bugs, can be a significant concern. To minimize the chance of these pests hitching a ride back with you, it's important to take specific precautions with your luggage.

Firstly, always keep your luggage away from the bed and other furniture in your hotel room. Bed bugs are notorious for hiding in mattresses, bed frames, and upholstered furniture, so placing your suitcase on these items can increase the risk of infestation. Instead, use the luggage rack provided by the hotel. If no luggage rack is available, the bathroom is a safer alternative since bed bugs are less likely to be found in tile and hard surfaces. Avoid placing your luggage directly on the bed or floor to reduce the chances of bed bugs climbing into your bags.

When packing, consider using bed bug-proof luggage liners or encasements. These products provide an extra layer of protection by preventing bed bugs from getting into your luggage. They are designed to be impenetrable to bed bugs, thus safeguarding your belongings. Additionally, store your clothing in sealed plastic bags or containers while traveling. This extra step can help minimize the risk of bed bugs contaminating your clothes and other personal items.

Another simple but effective measure is to keep your luggage zipped up when not in use. An open suitcase is an invitation for bed bugs to crawl inside, so keeping it closed can significantly reduce this risk. If you can, opt for hard-sided luggage over soft-sided options. Bed bugs are more likely to hide in the fabric and seams of soft luggage, whereas hard surfaces provide fewer hiding spots.

Post-Travel Inspection And Cleaning

Upon returning home from a trip, it is crucial to inspect and clean your belongings to prevent the introduction of bed bugs into your living space. Begin by unpacking your luggage outside your home or in a garage if possible. This initial step helps minimize the risk of bed bugs spreading within your residence.

Carefully examine all your clothing and other items before bringing them inside. Look for signs of bed bugs, such as small, reddish-brown insects or tiny black spots (their fecal matter). If you find any evidence of bed bugs, treat the affected items immediately. Wash your clothes in hot water and dry them on a high heat setting, as the heat will kill both bed bugs and their eggs. For items that cannot be washed, such as certain fabrics or delicate materials, consider placing them in a hot dryer for at least 30 minutes. Another effective method for these items is using a bed bug steam cleaner, which can penetrate fabrics and crevices to eliminate pests.

Next, vacuum your luggage thoroughly. Pay special attention to seams, zippers, and any crevices where bed bugs might hide. Use a vacuum with a high-efficiency particulate air (HEPA) filter if possible, as it will trap even the smallest particles. After vacuuming, immediately dispose of the vacuum bag in a sealed plastic bag to avoid any risk of bed bugs escaping and infesting your home.

Additionally, inspect and vacuum any areas where you stored your luggage during your trip. This includes closets, storage spaces, or any other places where your bags may have been placed. Vacuum these areas thoroughly, focusing on corners, baseboards, and other potential hiding spots for bed bugs. Again, dispose of the vacuum bag immediately after cleaning.

CHAPTER FIVE

EARLY DETECTION TECHNIQUES

Regular Home Inspections

Performing regular inspections of your home is crucial for early detection of bed bugs. These tiny pests are adept at hiding, making thorough inspections essential to catch them before an infestation takes hold.

Inspecting Sleeping Areas

Begin your inspection in the most common areas where bed bugs are likely to reside: beds, mattresses, and bedding. Carefully examine the seams, folds, and crevices of mattresses and box springs. Look for small, reddish-brown bugs, which are adult bed bugs, tiny white eggs, or black spots, which are bed bug feces. These signs are often found in clusters and can be a clear indication of an infestation.

Checking Furniture

Bed bugs are not limited to sleeping areas and can also infest furniture, particularly upholstered pieces. Inspect couches, chairs, and any other furniture near sleeping areas. Pay close attention to seams, under cushions, and in the frames. Bed bugs can easily migrate from beds to nearby furniture, making it essential to check these areas thoroughly.

Examining Other Hiding Spots

Bed bugs are notorious for hiding in less obvious places. Cracks and crevices in walls, baseboards, and electrical outlets are common hiding spots. They can also be found behind picture frames and even inside appliances. Use a flashlight and a magnifying glass to inspect these areas carefully. The small size of bed bugs makes them easy to miss, so taking the time to examine these hidden spots can make a significant difference in early detection.

Regular Schedule

Incorporating inspections into your regular housekeeping routine can help prevent a bed bug problem from escalating. Conduct weekly checks of your sleeping areas and furniture, and periodically inspect less obvious hiding spots. Regular inspections can catch early signs of bed bugs, allowing you to take action before the infestation becomes severe.

Using Bed Bug Interceptors And Monitors

Early detection of bed bugs is crucial for preventing a full-blown infestation in your home. One effective method for early detection involves using bed bug interceptors and monitors. These tools can help you identify the presence of bed bugs before they have a chance to spread.

Bed Bug Interceptors: These devices are small, plastic dishes placed under the legs of beds and other furniture. They operate by trapping bed bugs as they attempt to climb up to feed. The design of the interceptors ensures that once the bugs are caught in the dish, they cannot escape. To use interceptors effectively, place them under all the legs of your bed and any furniture that could be potential hiding spots for bed bugs. It's important to regularly check these interceptors for any trapped bed bugs. Finding bugs in the interceptors is a clear indication that you might have an infestation, and it's time to take further action.

Bed Bug Monitors: Monitors are specialized devices designed to attract and capture bed bugs. They often use heat and CO_2 to mimic a human host, drawing bed bugs out of their hiding places. Once the bugs enter the monitor, they are trapped and can be easily removed and disposed of. To maximize their effectiveness, place monitors in areas where you suspect bed bug activity. This could include under beds, near couches, and in other areas where bed bugs are likely to travel. Frequent checks of these monitors will help you stay ahead of a potential infestation.

Proper Placement: The placement of both interceptors and monitors is critical to their success. Ensure that interceptors are correctly positioned under the legs of furniture, and place monitors in strategic locations where bed bugs are likely to be found. Regular inspection of these devices can give you an early warning of bed bug activity, allowing you to address the issue before it escalates.

Recognizing Early Signs Of Infestation

Recognizing early signs of a bed bug infestation is crucial for taking prompt and effective action. Understanding these signs can help you address the problem before it becomes more severe and difficult to manage.

One of the initial indicators of bed bugs is often bite marks on your skin. Bed bug bites typically manifest as small, red, itchy welts, often arranged in a line or cluster. These bites can be uncomfortable and may cause allergic reactions in some people. However, it's important to note that not everyone reacts to bed bug bites. Therefore, the absence of bite marks does not necessarily mean that your home is free of bed bugs.

Another clear sign of a bed bug infestation is the presence of shed skins and eggs. As bed bugs grow, they shed their skins, leaving behind empty, translucent shells that resemble the bugs themselves. These shed skins are usually found in areas where bed bugs hide, such as mattress seams, furniture joints, and baseboards.

Additionally, bed bugs lay tiny, white, oval-shaped eggs, about the size of a pinhead. These eggs are often found in the same hiding spots and can be difficult to spot due to their small size and light color.

Unusual odors can also indicate a bed bug infestation. Bed bugs emit a musty, sweet smell, especially when they are present in large numbers. If you notice an unusual odor in your home, it could be a sign that bed bugs are nearby. This smell is often described as being similar to the scent of rotten raspberries or almonds.

Blood stains and fecal spots are additional signs to watch for. Blood stains can appear on your sheets, pillowcases, or mattress if you accidentally crush a bed bug after it has fed. These stains are usually small and may go unnoticed unless inspected closely.

Bed bug feces, on the other hand, appear as small, dark spots that can be found on bedding, mattresses, walls, or furniture. These spots are the digested blood left behind by the bugs.

If you identify any of these signs, it's advisable to contact a pest control professional. Professionals can confirm the presence of bed bugs and recommend the best course of action for treatment and prevention, ensuring that the infestation is effectively managed and eliminated.

CHAPTER SIX

SAFE AND EFFECTIVE TREATMENTS

Non-Chemical Treatment Options

Non-chemical treatments are often the first line of defense against bed bugs. These methods are generally safer for humans and pets and can be quite effective when done correctly.

Heat Treatment: Bed bugs are highly sensitive to heat. Washing bedding, clothing, and other infested items in hot water and drying them on the highest heat setting can kill bed bugs and their eggs. Additionally, steam cleaning can be used on mattresses, furniture, and other areas where bed bugs may hide. Steamers work well because they can penetrate deep into fabrics and crevices where bed bugs reside, reaching the high temperatures needed to exterminate both bugs and their eggs.

Cold Treatment: Exposure to freezing temperatures can also kill bed bugs. Items such as clothing, shoes, and bedding can be placed in plastic bags and stored in a freezer at temperatures below 0°F (-18°C) for several days. This method is effective for items that cannot be washed or steamed. Ensuring that the temperature remains consistently below freezing is crucial for this treatment to work effectively.

Vacuuming: Regular vacuuming of mattresses, furniture, carpets, and other areas can help remove bed bugs and their eggs. Be sure to empty the vacuum bag or canister outside the home immediately after vacuuming to prevent re-infestation.

Using a vacuum with a HEPA filter can enhance effectiveness by trapping even the smallest particles, including bed bug eggs.

Encasements: Using mattress and box spring encasements can trap bed bugs and prevent them from spreading. These encasements are designed to be impermeable to bed bugs and can be an effective preventive measure. By encasing the mattress and box spring, bed bugs are sealed inside and cannot escape, eventually dying from starvation.

Diatomaceous Earth: This natural, non-toxic powder can be sprinkled around bed frames, baseboards, and other areas where bed bugs are known to travel. Diatomaceous earth works by dehydrating and killing the bugs upon contact. When the powder adheres to the bed bugs, it abrades their exoskeletons, causing them to dry out and die. It is important to use food-grade diatomaceous earth and apply it carefully to avoid inhalation.

Chemical Treatments And Safety

When non-chemical methods prove insufficient in controlling bed bug infestations, chemical treatments can be employed. It is essential to use these treatments safely and adhere to the manufacturer's instructions to avoid potential health risks.

Insecticides: Several insecticides are specifically designed for bed bug control. These include pyrethrins, pyrethroids, and neonicotinoids. Pyrethrins are derived from chrysanthemum flowers and are effective against bed bugs by attacking their nervous systems. Pyrethroids are synthetic chemicals similar to pyrethrins but are longer-lasting. Neonicotinoids, another class of insecticides, act on the nervous system of insects, causing paralysis and death. When choosing insecticides, ensure the products are EPA-approved for bed bug control. Following the application guidelines carefully is crucial to ensure effectiveness and safety. Misapplication can lead to health hazards and may reduce the treatment's efficacy.

Residual Sprays: These sprays remain effective for an extended period after application, making them ideal for long-term bed bug control. Residual sprays can be applied to surfaces where bed bugs are likely to travel, such as baseboards, cracks, and crevices. The active ingredients in these sprays continue to kill bed bugs long after the initial application. However, it is vital to ensure treated areas are adequately ventilated. Following safety precautions, such as wearing protective gear and keeping pets and children away from treated areas, is essential to prevent potential health risks.

Dusts: Insecticidal dusts, such as silica gel and boric acid, are effective for treating voids, cracks, and other hard-to-reach areas. These dusts adhere to the bed bugs, causing them to dehydrate and die. Using a hand duster, apply the dust evenly to avoid creating airborne particles, which can be harmful if inhaled. Dust treatments are beneficial because they provide long-lasting protection and can reach areas that sprays might miss.

Aerosol Sprays: Aerosol sprays can be effective for spot treatments in small areas where bed bugs are present. These sprays typically contain pyrethrins or other fast-acting chemicals that kill bed bugs on contact. When using aerosols, exercise caution as they can be harmful if inhaled or used improperly. Ensure the area is well-ventilated during and after application, and follow the manufacturer's instructions to minimize health risks.

Professional Pest Control Services

When dealing with a severe pest infestation or when DIY methods fall short, enlisting professional pest control services can be essential. These experts offer advanced treatment options and possess the knowledge necessary to manage bed bug infestations both safely and effectively.

A key first step is a thorough inspection conducted by a pest control professional. This comprehensive assessment identifies the infestation's extent and pinpoints specific areas where bed bugs are concealing themselves. Such detailed inspections are crucial for formulating an effective treatment plan.

Many pest control services employ Integrated Pest Management (IPM) strategies. IPM is a holistic approach combining various treatment methods to achieve thorough bed bug control. This approach integrates chemical and non-chemical treatments, environmental adjustments, and ongoing monitoring to tackle the problem from multiple angles.

One of the most effective methods used by professionals is heat treatment.

Professional-grade heat treatments surpass the capabilities of DIY methods by achieving higher temperatures and covering broader areas. Specialized equipment is used to heat entire rooms to temperatures lethal to bed bugs and their eggs, ensuring a thorough eradication.

In cases where infestations are particularly severe, fumigation might be necessary. This method involves sealing off the affected area and introducing a fumigant gas that permeates all hiding spots, eliminating bed bugs throughout the space. Due to its complexity and potential hazards, fumigation should only be performed by licensed professionals.

Follow-up treatments are a standard component of professional pest control services. These subsequent visits ensure that the infestation has been completely eradicated. Continuous monitoring and preventive measures are crucial to avoiding re-infestation and maintaining a pest-free environment.

CHAPTER SEVEN

LONG-TERM BED BUG MANAGEMENT

Developing A Bed Bug Prevention Plan

To effectively manage and prevent bed bugs, it's essential to establish a comprehensive prevention plan that encompasses both immediate actions and ongoing practices. Begin by conducting a thorough inspection of your home to identify potential hiding spots. Bed bugs are adept at concealing themselves in small spaces, so scrutinize areas such as cracks, crevices, and under furniture meticulously. Look for signs of infestation, such as small reddish-brown stains, shed skins, or live insects.

Once you have identified potential problem areas, the next step is to seal any cracks or crevices in walls, floors, and around windows and doors. Bed bugs can enter through even the tiniest openings, so thorough sealing is crucial to prevent their return. Use caulk or other appropriate sealing materials to address these vulnerabilities. Regular maintenance checks and repairs will help ensure that these entry points remain secure.

In addition to sealing entry points, regular vacuuming of carpets, rugs, and upholstery is a key preventive measure. Vacuuming helps remove any bed bugs, eggs, or larvae that may be present. After vacuuming, dispose of the vacuum bag immediately and outside of your home to avoid the risk of reintroducing bed bugs.

Another effective strategy is the use of mattress and box spring encasements. These encasements are specially designed to trap bed bugs and their eggs, preventing them from infesting your mattress or box spring. Choose encasements that are labeled as bed bug-proof and ensure they are securely zipped to avoid any breaches. Regularly inspect these encasements for any signs of damage or wear.

Educating Family And Household Members

Educating your family and household members is essential for effective long-term management of bed bugs. Bed bugs are notorious for spreading quickly if not properly managed, and ensuring that everyone in the home is informed can make a significant difference in controlling and preventing infestations.

Begin by educating your household about the common signs of a bed bug infestation. Bed bugs are small, reddish-brown insects that are often difficult to spot with the naked eye. Look out for signs such as bite marks on the skin, which may appear in clusters or rows and can cause itching and discomfort. Also, keep an eye out for small blood spots on bedding or sheets, which are indicative of bed bugs feeding on blood. Additionally, check for the presence of tiny, dark spots on mattresses and bed frames, which are bed bug excrement.

It is crucial that all members of the household understand the importance of reporting any potential bed bug sightings or symptoms immediately. Early detection is key to preventing a small issue from escalating into a larger, more difficult problem. Encourage a proactive approach by fostering a sense of vigilance, especially when traveling or staying in hotels. Bed bugs are notorious hitchhikers and can easily be transported from one location to another via luggage or clothing.

Teach your family members proper inspection and handling techniques to minimize the risk of bringing bed bugs into the home. For instance, when returning from a trip, thoroughly inspect all luggage and clothing for any signs of bed bugs before bringing them indoors. It is also advisable to wash and dry clothes and other fabric items at high temperatures, as this can kill any bed bugs or eggs that might be present.

Regular Follow-Up And Monitoring

Long-term bed bug management relies heavily on consistent follow-up and monitoring. Once you've implemented a prevention plan to address a bed bug infestation, it's crucial to maintain vigilance to ensure the problem doesn't resurface. Schedule regular inspections of your home to check for any new signs of bed bugs. Focus particularly on areas where these pests are known to hide, such as behind headboards, in cracks and crevices, and around baseboards.

To enhance your monitoring efforts, consider using bed bug detection tools. Bed bug monitors and traps can be highly effective in detecting these pests early, allowing you to address any issues before they escalate into significant problems.

Strategically place these detection tools in critical locations, such as near sleeping areas and potential entry points, to maximize their effectiveness.

During your inspections, thoroughly check all potential hiding spots for any signs of bed bugs, including live bugs, shed skins, or fecal stains. If you find any evidence of bed bugs, take immediate action. This may involve using targeted insecticides designed for bed bugs, enlisting professional pest control services, or reassessing and adjusting your prevention strategies.

It's important to stay proactive and not become complacent. Bed bugs are notoriously persistent and can be difficult to completely eradicate. Regular monitoring and prompt action are essential to ensuring that any new infestations are caught early and managed effectively. By maintaining a vigilant approach, you can minimize the risk of a full-blown infestation and keep your home bed bug-free. Remember, consistent follow-up is key to long-term success in managing and preventing bed bug problems.

THE END

www.ingramcontent.com/pod-product-compliance
Lightning Source LLC
Chambersburg PA
CBHW071843210526
45479CB00001B/262